P9-AOI-138

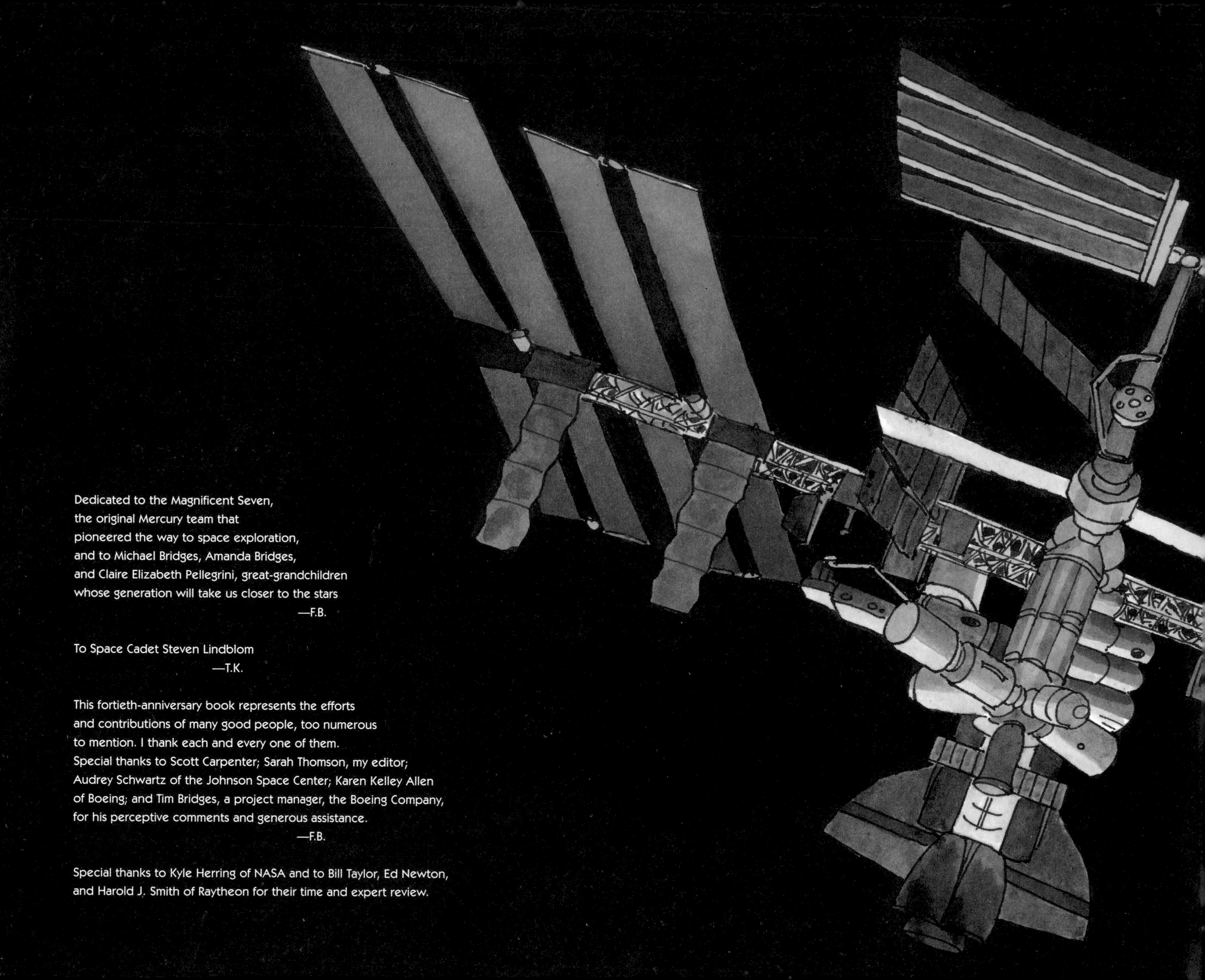

Dedicated to the Magnificent Seven,
the original Mercury team that
pioneered the way to space exploration,
and to Michael Bridges, Amanda Bridges,
and Claire Elizabeth Pellegrini, great-grandchildren
whose generation will take us closer to the stars
—F.B.

To Space Cadet Steven Lindblom
—T.K.

This fortieth-anniversary book represents the efforts
and contributions of many good people, too numerous
to mention. I thank each and every one of them.
Special thanks to Scott Carpenter; Sarah Thomson, my editor;
Audrey Schwartz of the Johnson Space Center; Karen Kelley Allen
of Boeing; and Tim Bridges, a project manager, the Boeing Company,
for his perceptive comments and generous assistance.
—F.B.

Special thanks to Kyle Herring of NASA and to Bill Taylor, Ed Newton,
and Harold J. Smith of Raytheon for their time and expert review.

LET'S-READ-AND-FIND-OUT SCIENCE®

The International Space Station

by Franklyn M. Branley

illustrated by True Kelley

GEMINI

Virgil L. Grissom and John W. Young
in orbit 4 hours and 53 minutes
1965

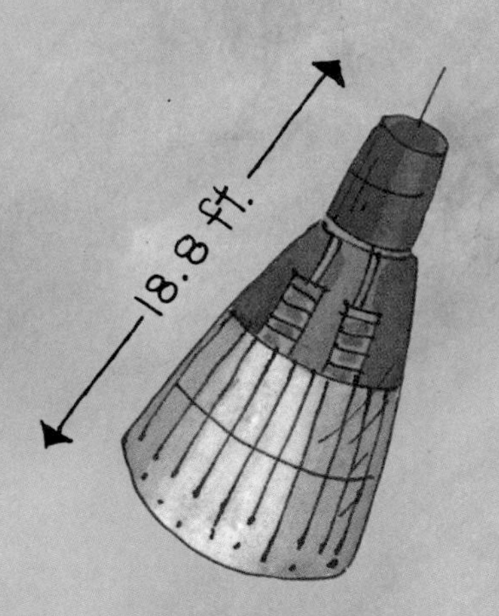

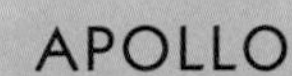

APOLLO

Neil Armstrong, Edwin Aldrin,
and Michael Collins
first lunar landing
1969

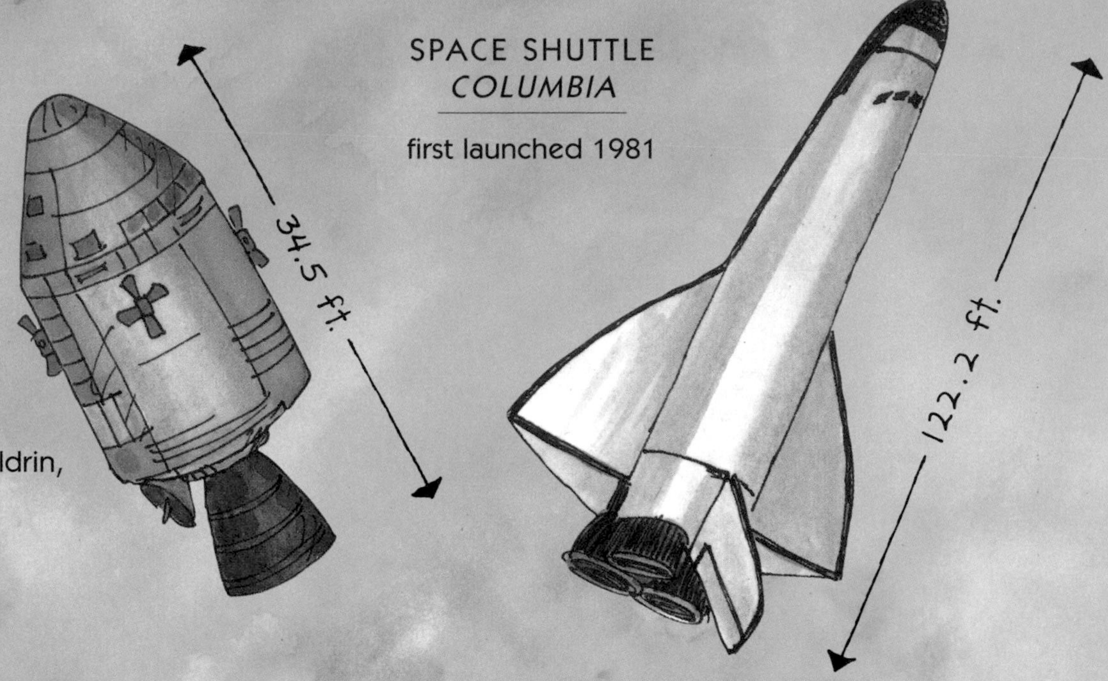

SPACE SHUTTLE
COLUMBIA

first launched 1981

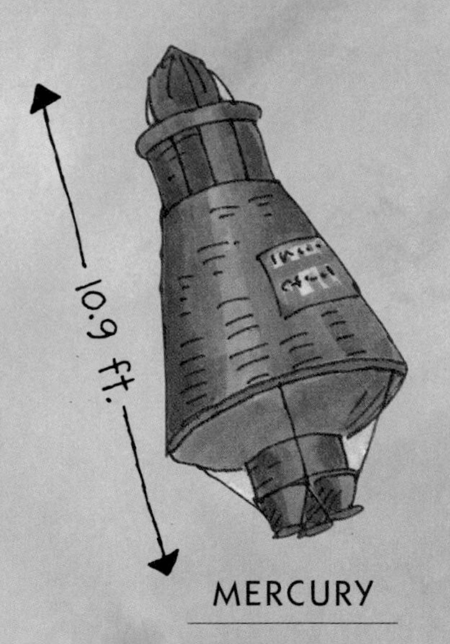

MERCURY

Scott Carpenter
in orbit 4 hours and 56 minutes
1962

For thousands of years, people have wondered what the moon is made of. We found the answer in 1969 when astronauts landed there. But before we could fly all the way to the moon, we had to learn to fly in orbit around Earth. This was the goal of the Mercury program, and it succeeded. In 1962, John Glenn was the first person to orbit Earth. I was the second.

It took many men and women many years to learn the thousands of things we needed to know before we could reach the moon. But we finally got there. We had taken our first giant step into outer space.

Now we are about to take another big step, with the International Space Station. In the Mercury capsule, I was in orbit four hours, all by myself. The Space Station will go around Earth for years, and teams of six or seven people will live and work there for many months. They will do research to find out the thousands of things we need to know before we can explore our solar system and the galaxy beyond.

One day, the Space Station will become the first step toward a mission to Mars. You may be a part of that mission.

How lucky you are!

Scott Carpenter

MERCURY ASTRONAUT SCOTT CARPENTER

RUSSIAN SPACE STATION MIR

1986

In December of 1998 two astronauts floating in outer space fastened together the first two parts of the International Space Station. The Russian part was called *Zarya*, which means "sunrise." The American part was called *Unity*. Together, *Zarya* and *Unity* were as tall as a seven-story building and weighed 77,000 pounds.

But that was just the beginning.

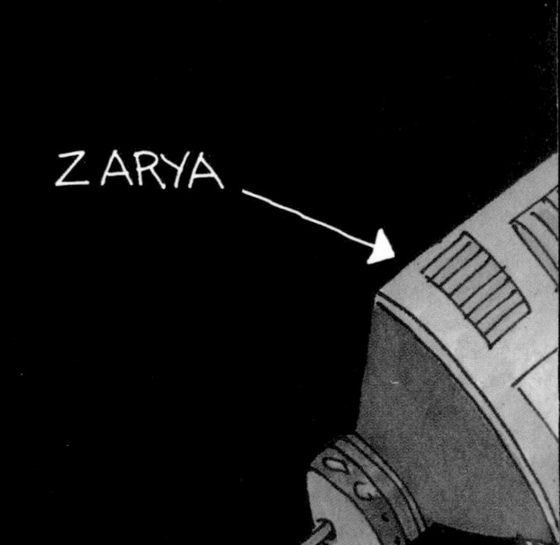

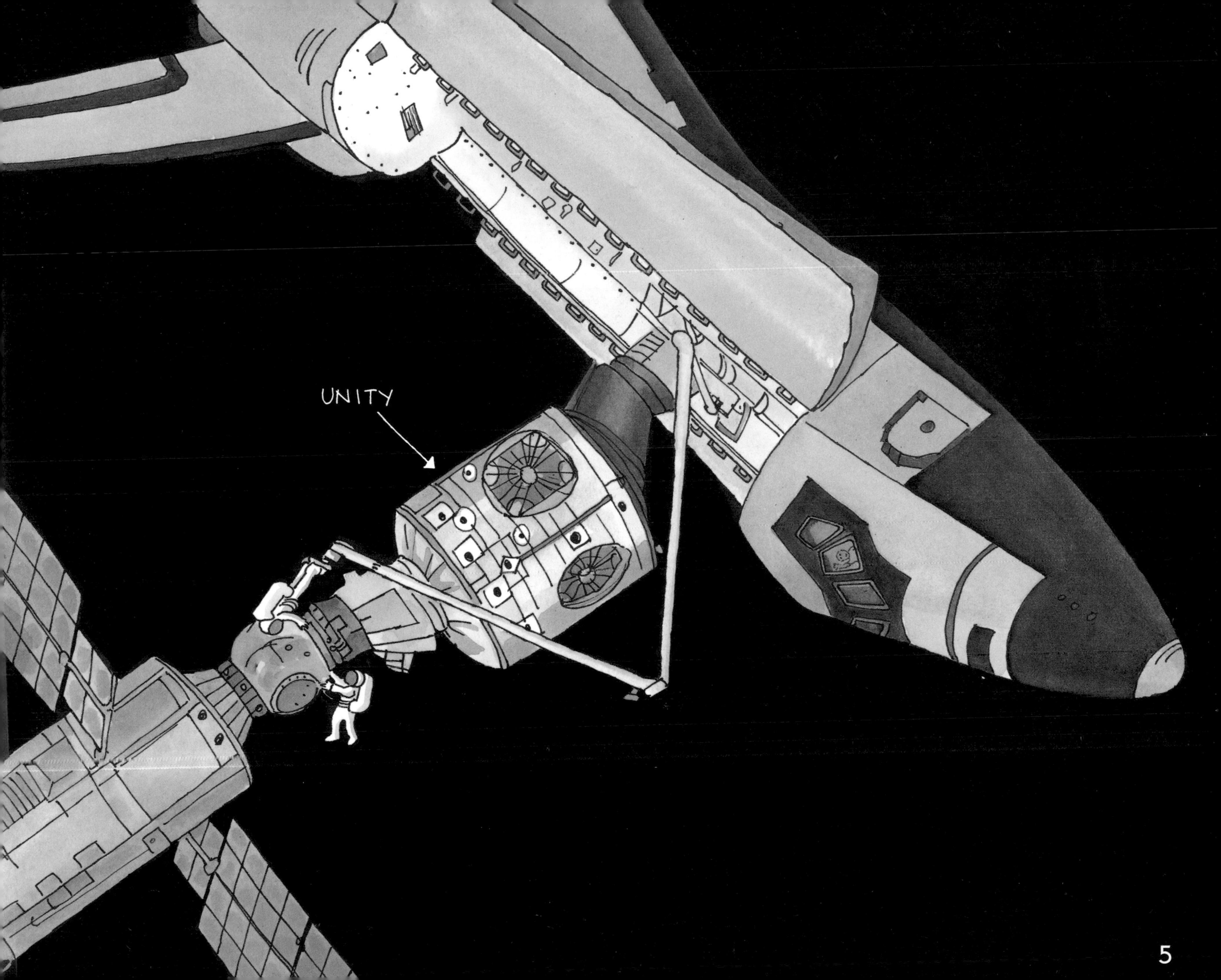
UNITY

The International Space Station is still being built today. It has thousands of parts. Some of them are capsules, or cabins that are fastened together. There are capsules called habitats where the crew live. There are laboratory and experiment capsules where they work, and capsules for storage, communications, and more.

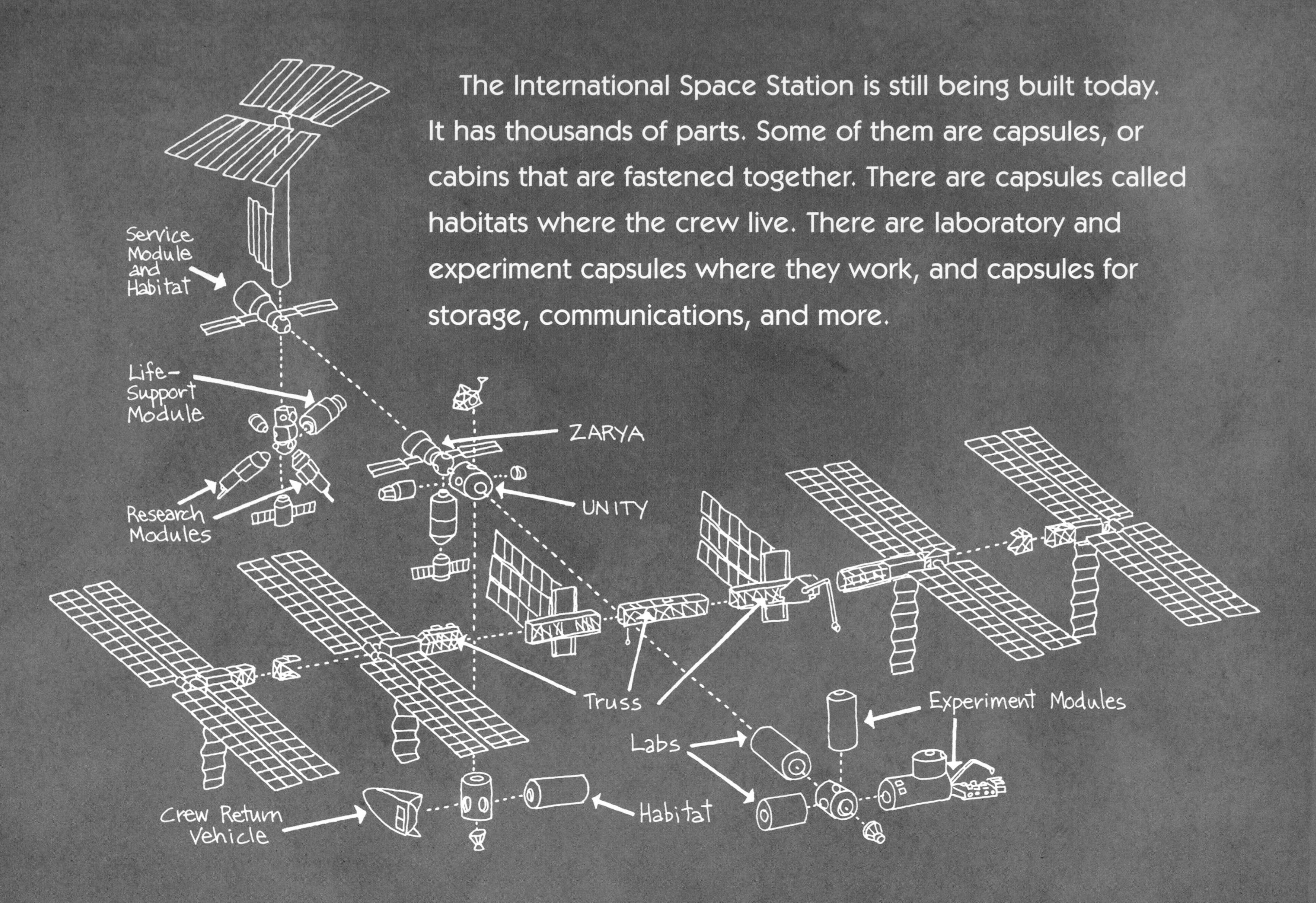

The backbone of the Station is a long piece called a truss. It holds everything together. Astronauts use the truss to move from one area to another.

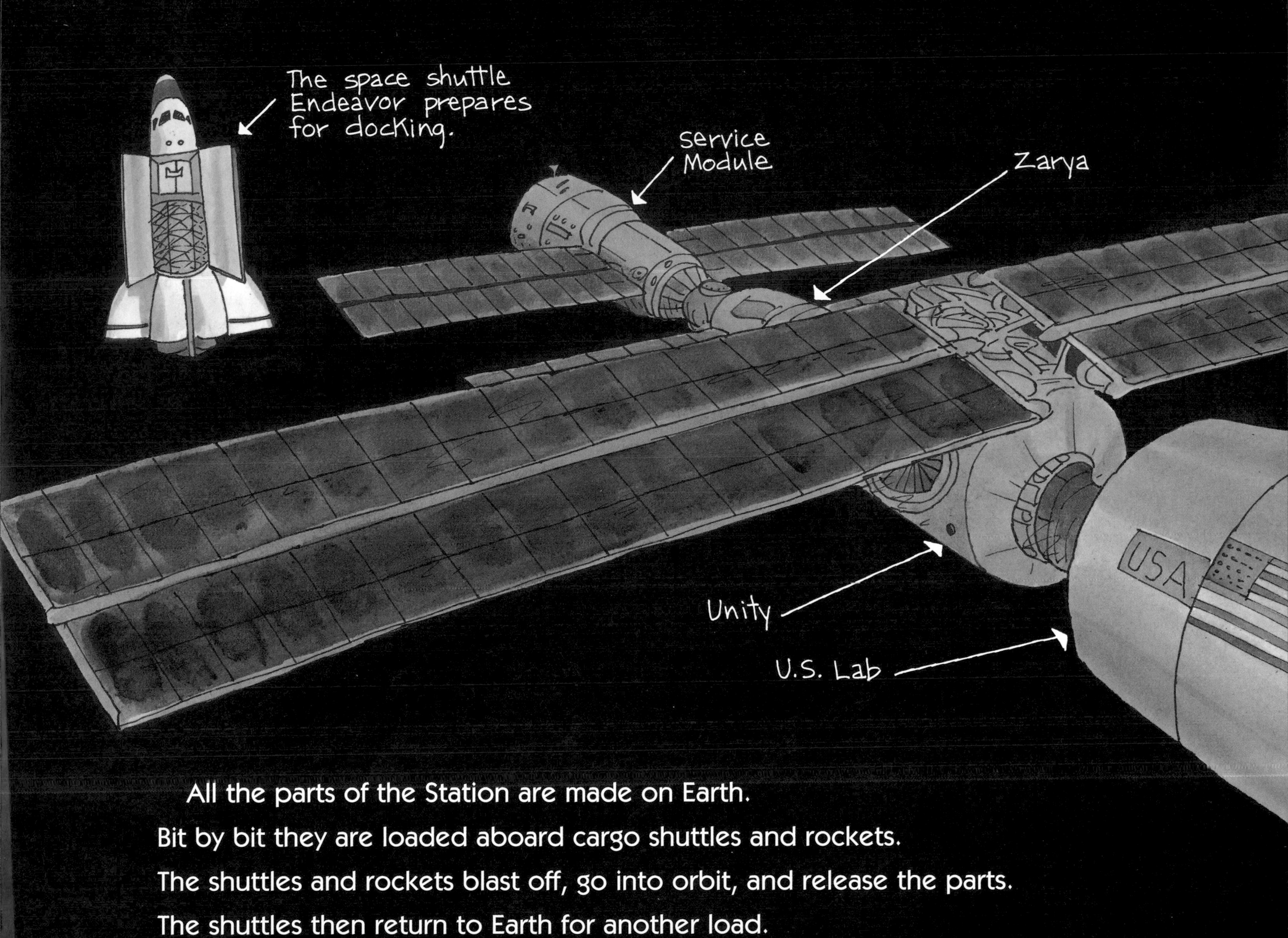

All the parts of the Station are made on Earth.
Bit by bit they are loaded aboard cargo shuttles and rockets.
The shuttles and rockets blast off, go into orbit, and release the parts.
The shuttles then return to Earth for another load.

After the parts of the Station are in orbit, they are put together. That's the special job of spacewalkers, men and women who have been trained to work where there is no air or gravity. Robots do some of the work outside the Station. But spacewalkers do most of it.

Spacewalkers must wear space suits to keep them from freezing—and from boiling. The suit also has air for the astronauts to breathe. A person can stay in a space suit for eight hours.

The suit keeps pressure pushing on the astronaut. Here on Earth air pushes in on our bodies all the time. But we don't feel it, because there is air inside our bodies pushing out. In outer space there is no air pushing in. But air inside a person still pushes out. So without a space suit, the body would swell up. It would push out farther and farther—and the astronaut would burst.

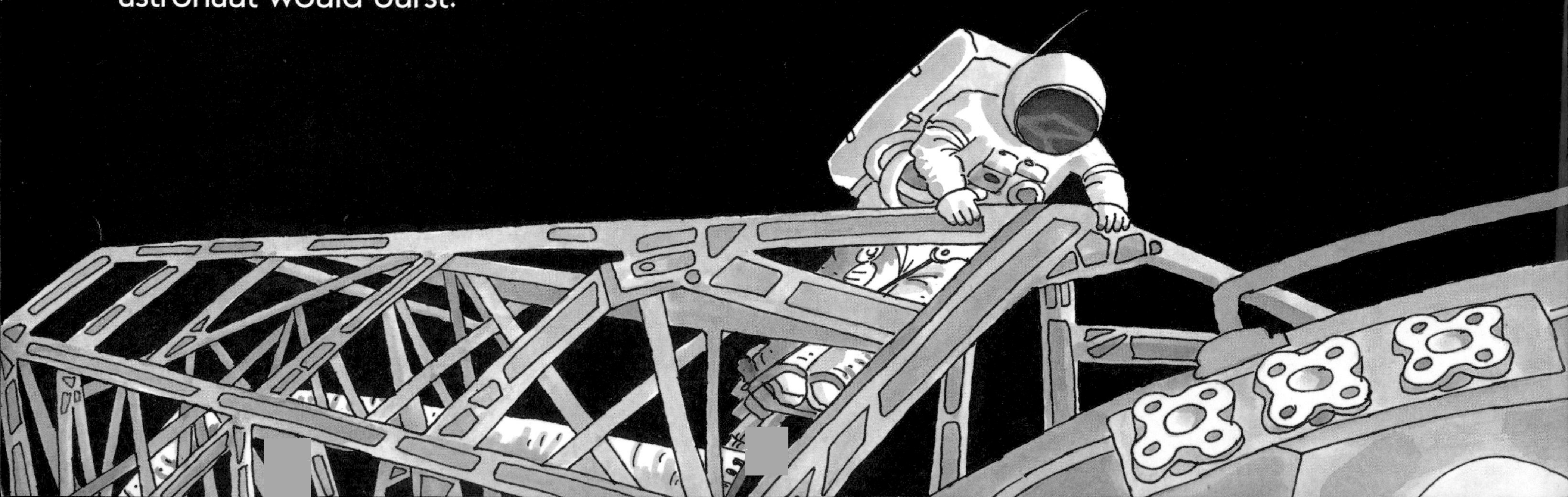

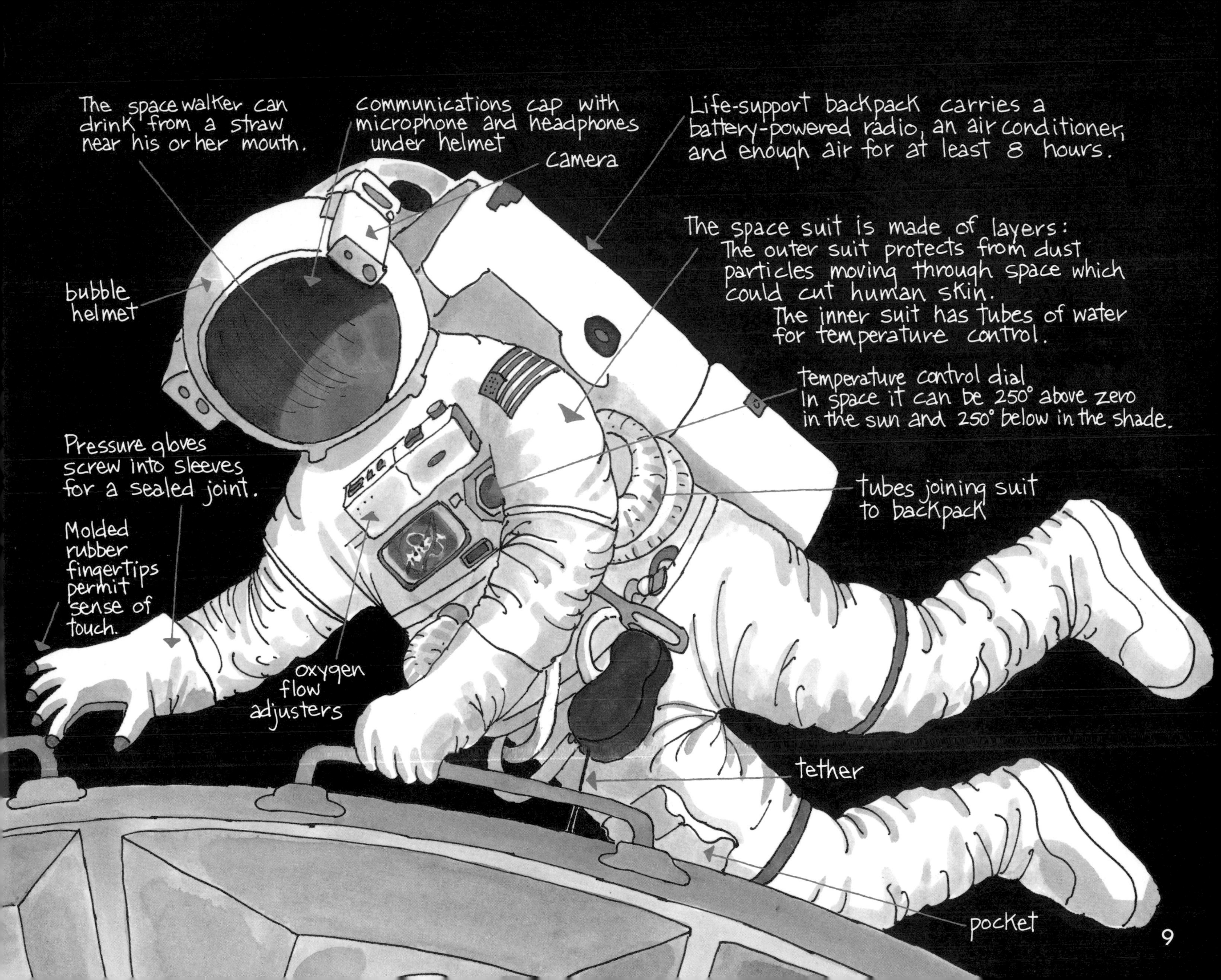
The spacewalker can drink from a straw near his or her mouth.
Communications cap with microphone and headphones under helmet
camera
Life-support backpack carries a battery-powered radio, an air conditioner, and enough air for at least 8 hours.
bubble helmet
The space suit is made of layers: The outer suit protects from dust particles moving through space which could cut human skin. The inner suit has tubes of water for temperature control.
temperature control dial
In space it can be 250° above zero in the sun and 250° below in the shade.
Pressure gloves screw into sleeves for a sealed joint.
Molded rubber fingertips permit sense of touch.
tubes joining suit to backpack
oxygen flow adjusters
tether
pocket

Spacewalkers are tied to the Station by a tether. Tools used by the spacewalkers must also be tethered. If they weren't, they would drift away and be lost.

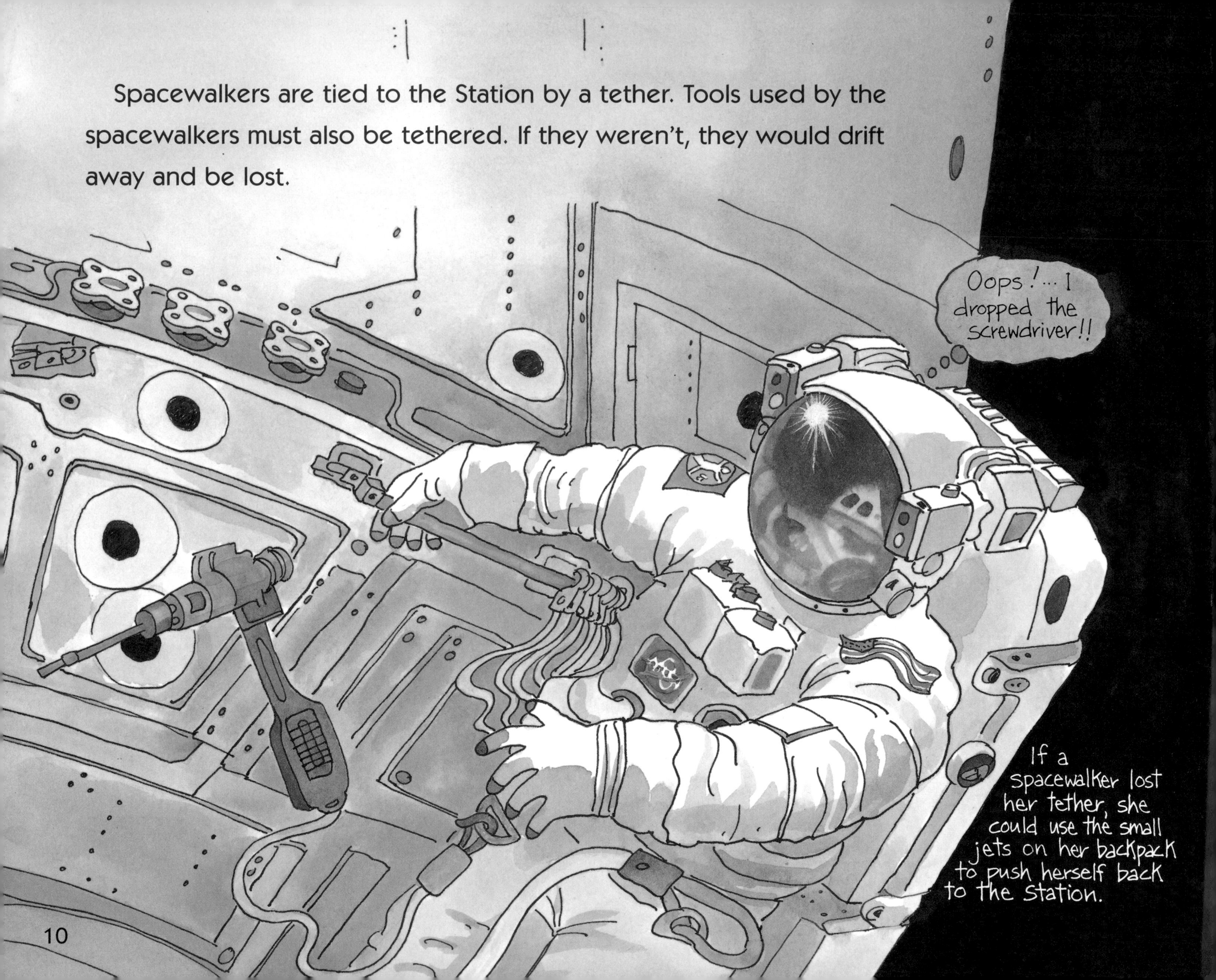

Spacewalkers move by pulling themselves from one handhold or foothold to another. They have to put their feet in footholds when they use screwdrivers or wrenches. If they didn't, it is the spacewalkers who would spin around instead of the screws or the nuts they were trying to tighten.

In several years the International Space Station will be finished.

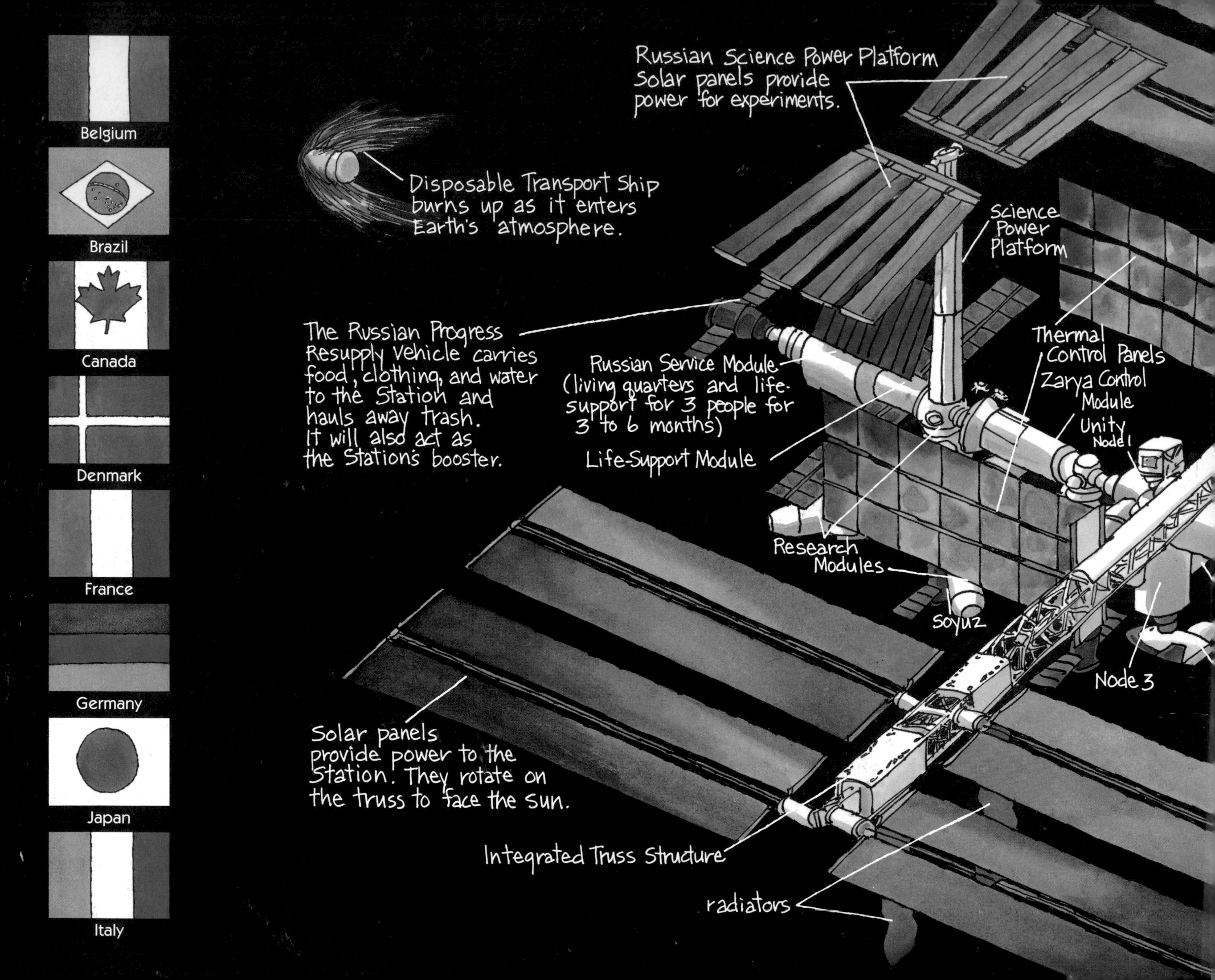
Belgium
Brazil
Canada
Denmark
France
Germany
Japan
Italy
Russian Science Power Platform
Solar panels provide
power for experiments.
Disposable Transport Ship
burns up as it enters
Earth's atmosphere.
Science
Power
Platform
The Russian Progress
Resupply Vehicle carries
food, clothing, and water
to the Station and
hauls away trash.
It will also act as
the Station's booster.
Russian Service Module
(living quarters and life-
support for 3 people for
3 to 6 months)
Life-Support Module
Thermal
Control Panels
Zarya Control
Module
Unity
Node 1
Research
Modules
Soyuz
Node 3
Solar panels
provide power to the
Station. They rotate on
the truss to face the sun.
Integrated Truss Structure
radiators

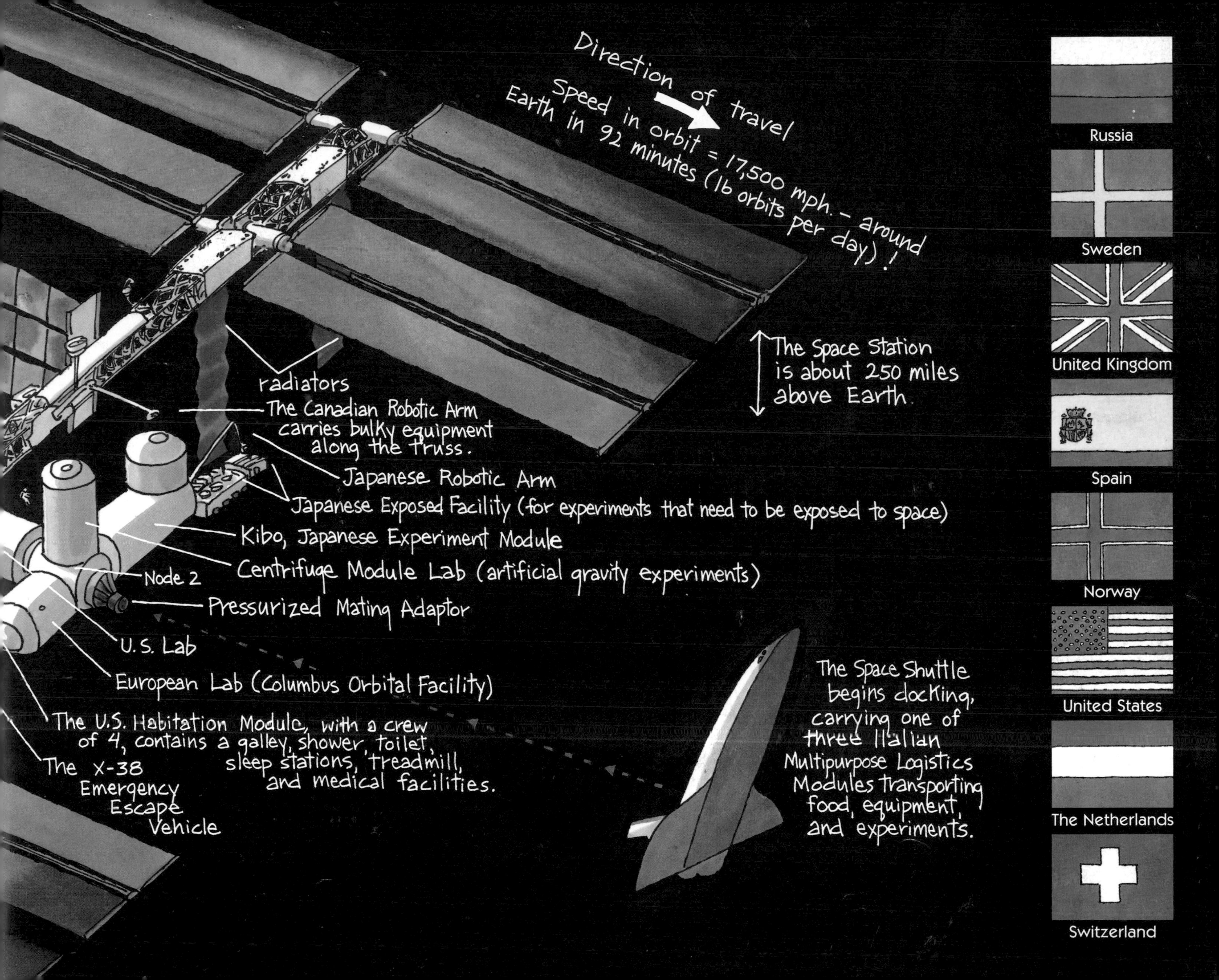

Direction of travel
Speed in orbit = 17,500 mph. – around Earth in 92 minutes (16 orbits per day)!
radiators
The Canadian Robotic Arm carries bulky equipment along the truss.
Japanese Robotic Arm
Japanese Exposed Facility (for experiments that need to be exposed to space)
Kibo, Japanese Experiment Module
Centrifuge Module Lab (artificial gravity experiments)
Node 2
Pressurized Mating Adaptor
U.S. Lab
European Lab (Columbus Orbital Facility)
The U.S. Habitation Module, with a crew of 4, contains a galley, shower, toilet, sleep stations, treadmill, and medical facilities.
The X-38 Emergency Escape Vehicle
The Space Station is about 250 miles above Earth.
The Space Shuttle begins docking, carrying one of three Italian Multipurpose Logistics Modules transporting food, equipment, and experiments.
Russia
Sweden
United Kingdom
Spain
Norway
United States
The Netherlands
Switzerland

The Station will be more than 350 feet long—as long as a 30-story building is tall. It will weigh about a million pounds. It seems as if something that heavy would fall down—and actually, the Station does! As the Station circles Earth, it slowly gets closer and closer to the planet. If it weren't boosted back up, the Station would crash into Earth.

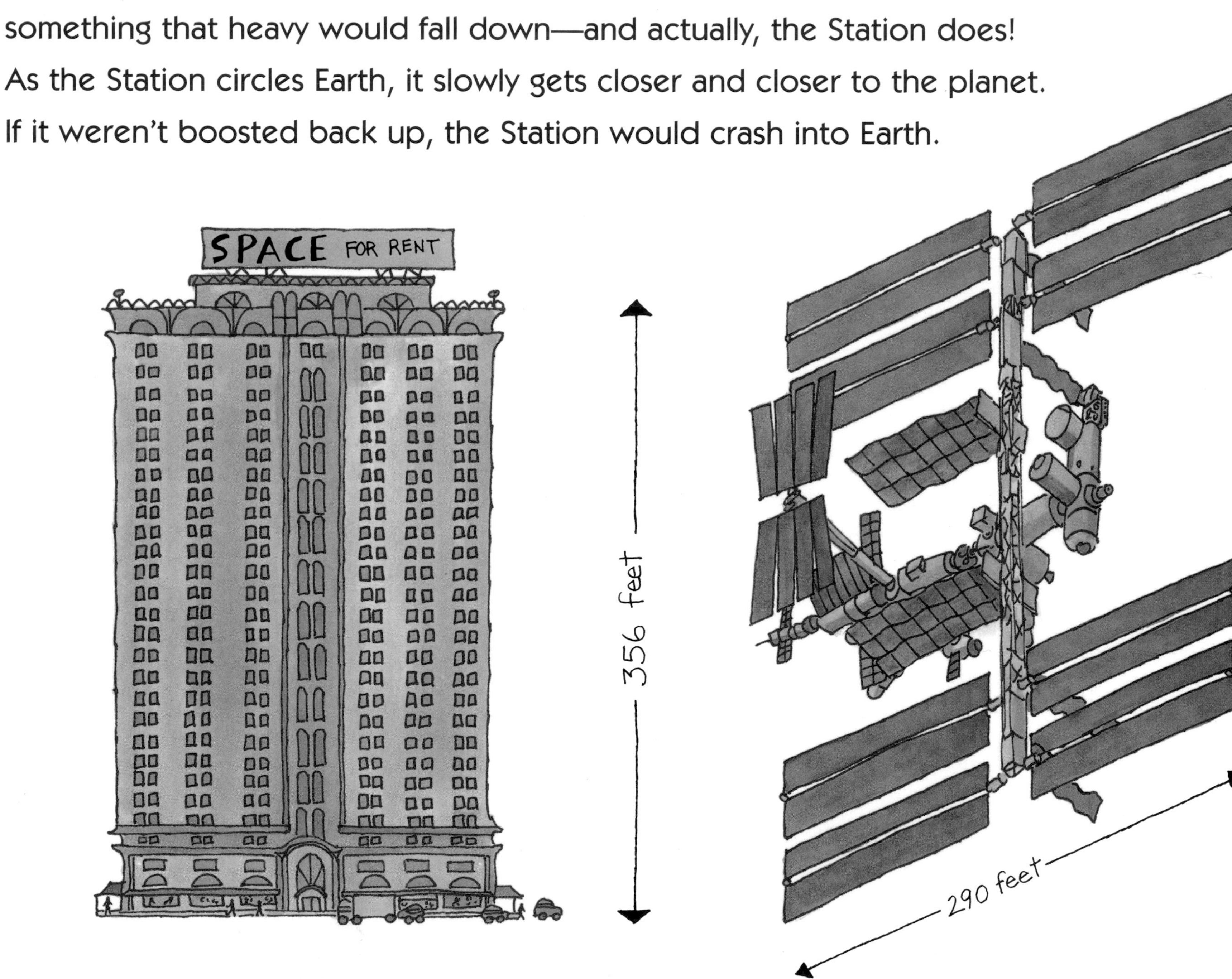

That's why there are small engines on the Station. They help push the Station back into a higher orbit when it slows down and moves closer to Earth.

Almost everything on the Station is run by computers. Those computers use a lot of electricity, and it all has to be made on board. The large flat sections at both ends of the Station are the solar panels. They make the electricity that runs the Station.

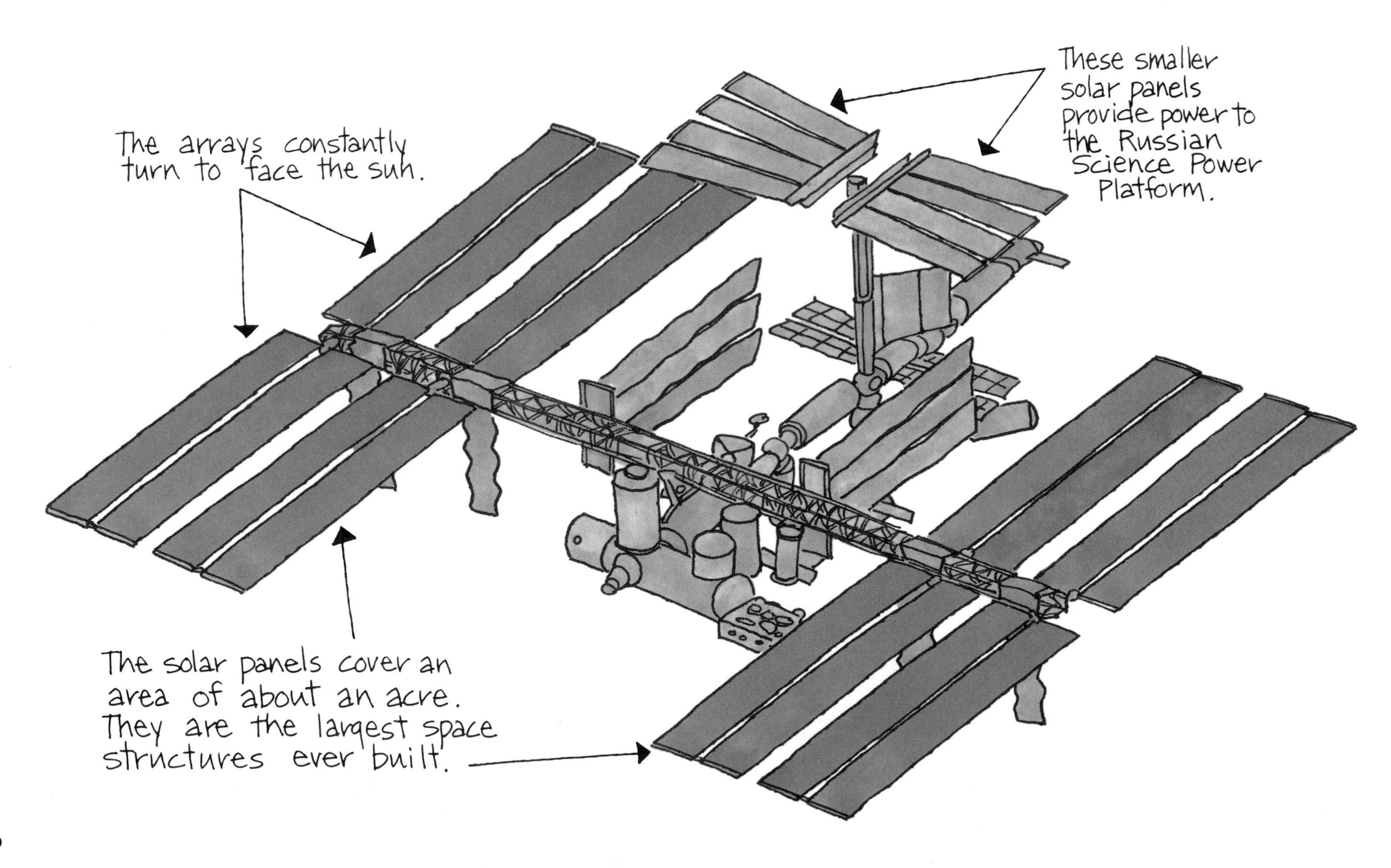

Solar cells cover the solar panels. The cells change the energy of sunlight into electricity. There are never any clouds in space, so unless Earth is between the Station and the sun, there is always sunlight.

When the Station is in Earth's shadow, lots of batteries are used for electricity. They are charged while the ship is in sunshine.

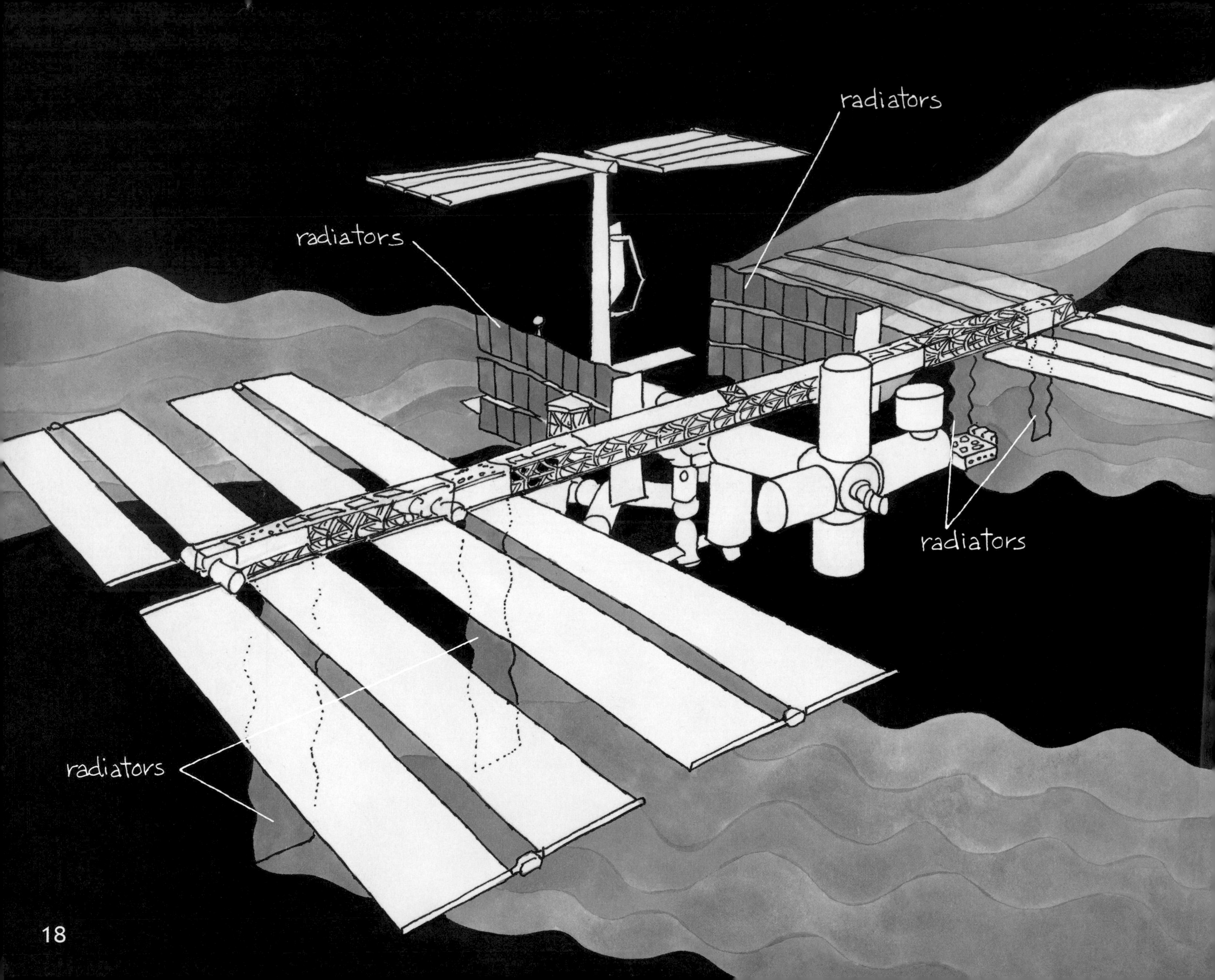
radiators
radiators
radiators
radiators

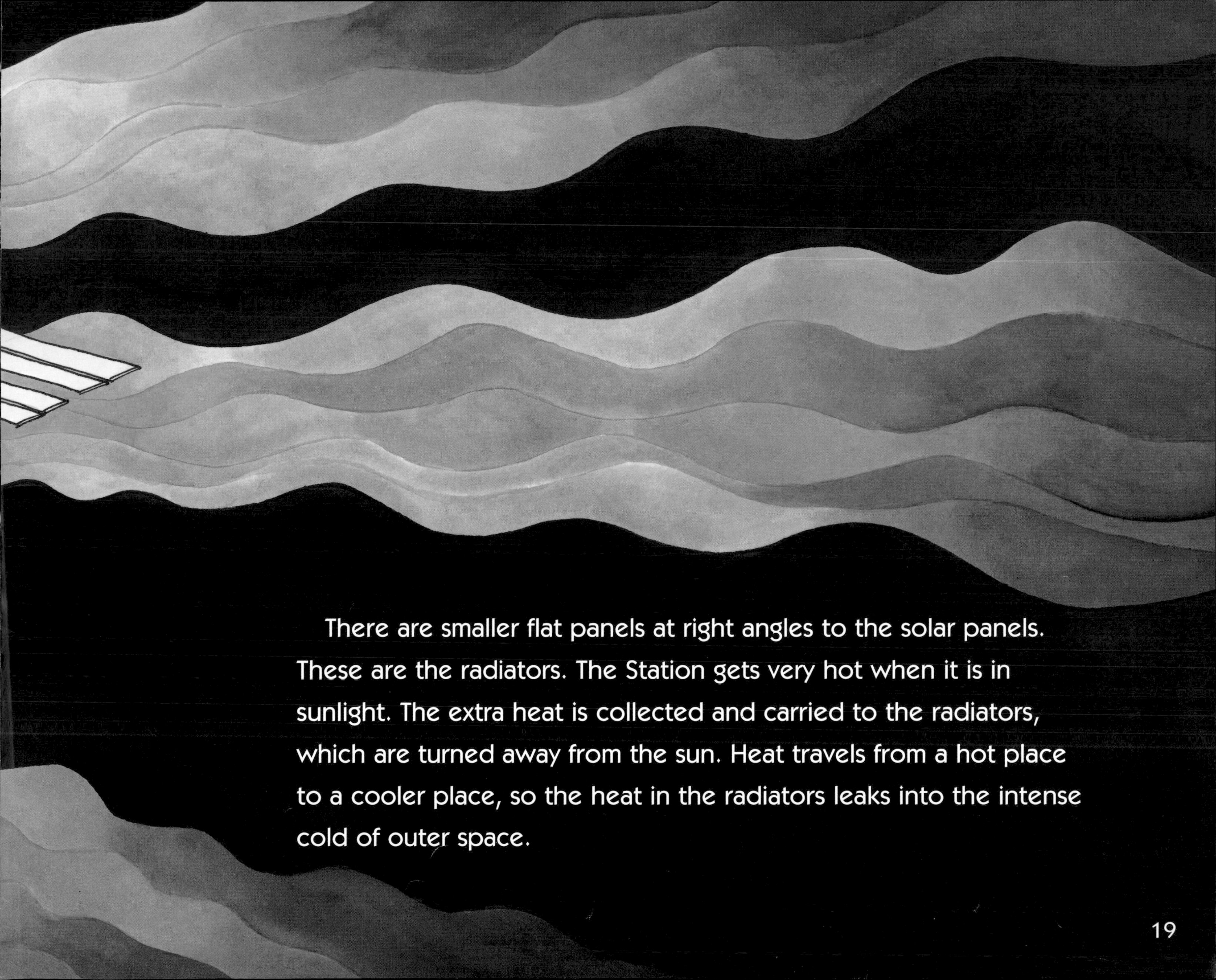

There are smaller flat panels at right angles to the solar panels. These are the radiators. The Station gets very hot when it is in sunlight. The extra heat is collected and carried to the radiators, which are turned away from the sun. Heat travels from a hot place to a cooler place, so the heat in the radiators leaks into the intense cold of outer space.

What if there were trouble on board the Station? The Station might be hit by a meteorite, or something could go wrong with one of the computers. That's why there will be an escape vehicle called the X-38 on the Station. It looks a little like a high-speed boat, and it could carry all the astronauts safely back to Earth.

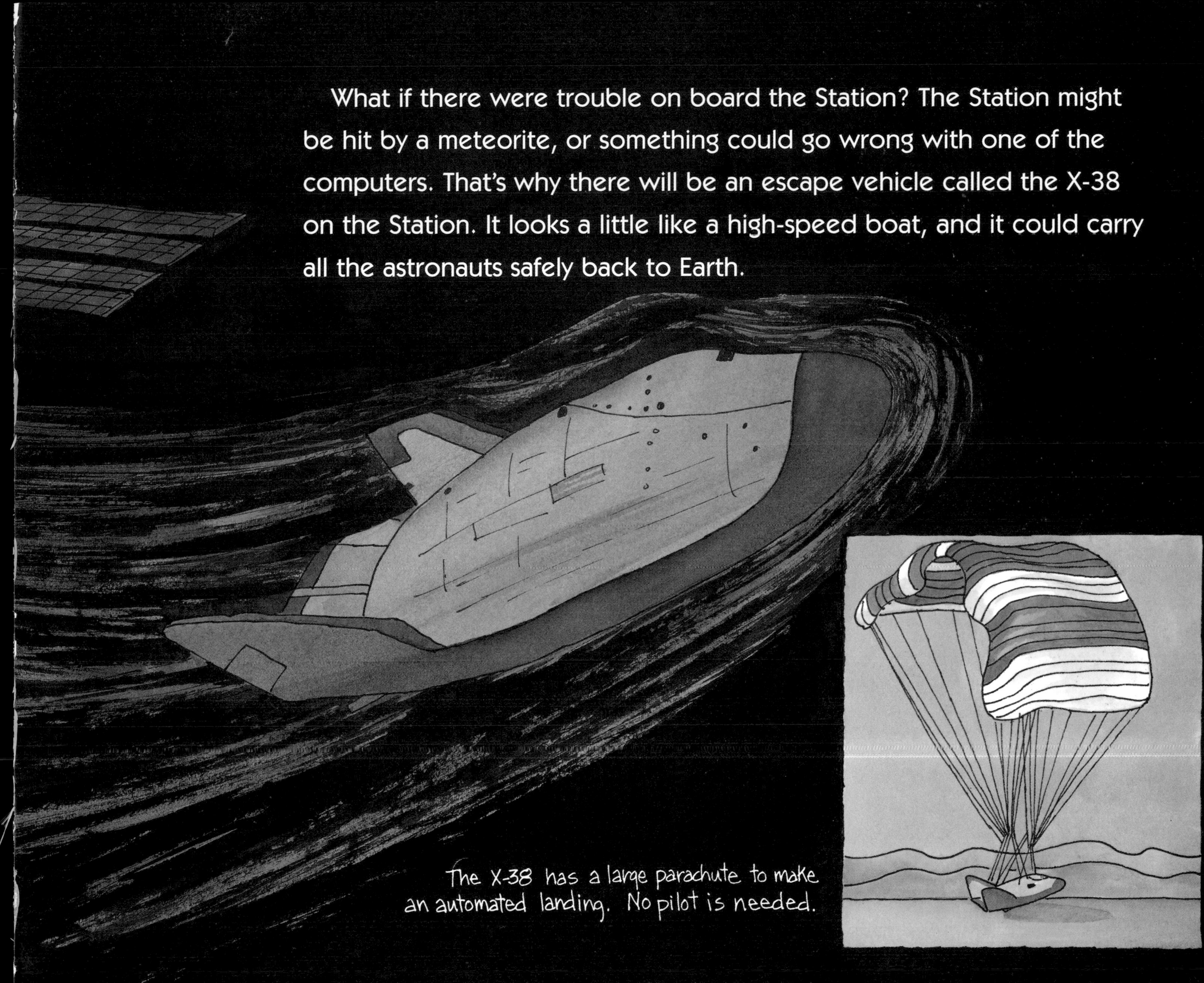

The X-38 has a large parachute to make an automated landing. No pilot is needed.

Men and women from the United States, Japan, England, and Russia—and eventually from sixteen other countries—live and work on the Station. What do they do?

The U.S. Habitation Module

The first thing they do is try to stay healthy.

There is air inside the Station, so the astronauts don't have to wear their space suits. But the crew is still in outer space. They are in zero gravity—or very close to it. They float in space. Their muscles and bones don't have to work very hard as they stand, sit, or move about. So their muscles and bones get weaker.

The crew must exercise a lot to use their muscles. Usually they use machines. They also eat good food. All of it is sent to them from Earth. A lot of the food is dried, and water is added to it at mealtime. However, the crew of the Station also has fresh fruit and vegetables brought by the shuttle. No one sits down for a meal. In place of chairs, foot holders keep the diners from floating away.

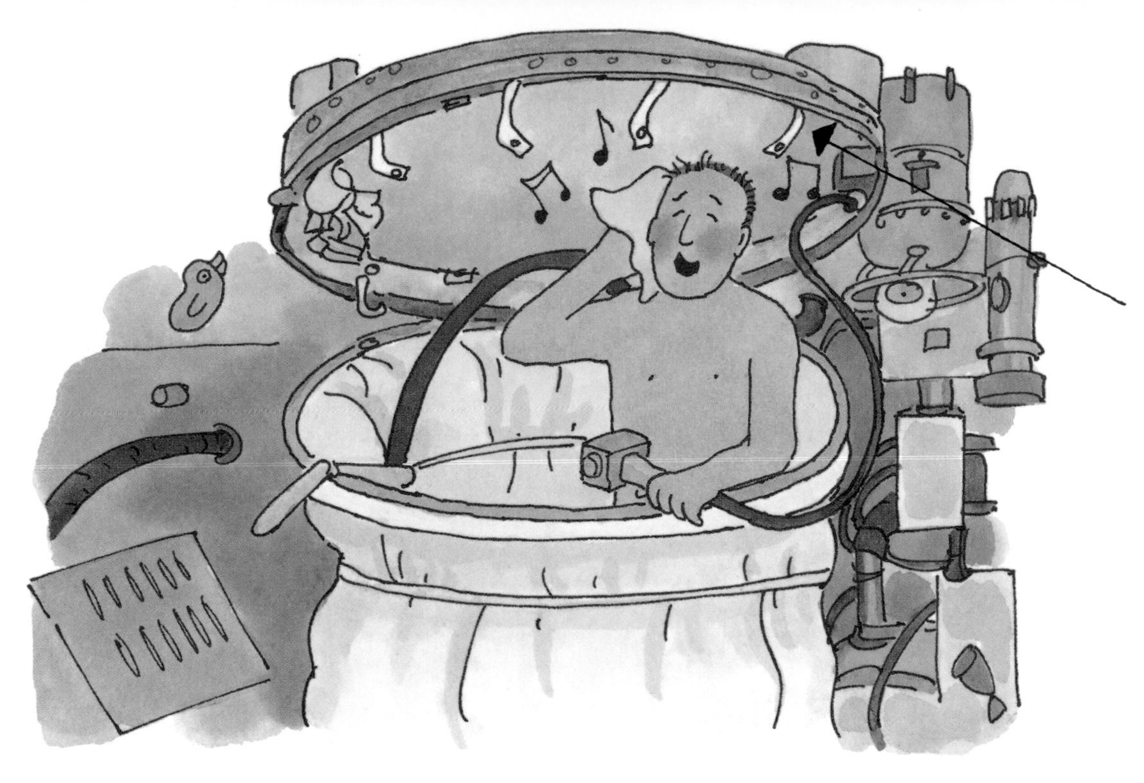

The shower has a cover to keep water droplets from floating around. They could be dangerous if they got into the electronic computer equipment.

Water doesn't flow when there is no gravity. So when the astronauts take a shower, water from a hose is rubbed on like a lotion. When it's time for bed, the astronauts crawl into anchored sleeping bags so they don't float away while they're sleeping.

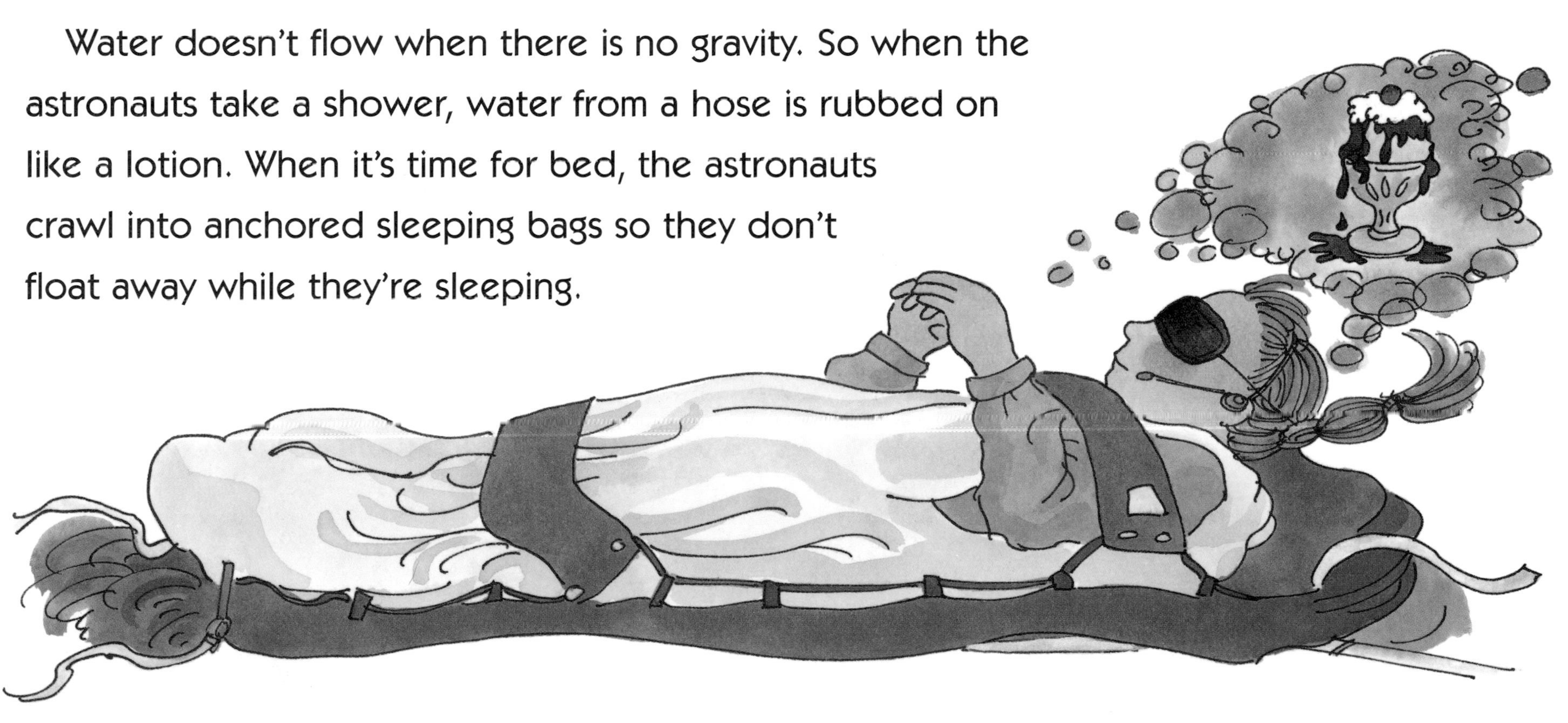

The astronauts on the Station do research to find out more about outer space. They can grow perfect crystals to study back on Earth. You can't grow perfect crystals on Earth—only in space. They can make alloys—mixtures of metals—that are better and stronger than the ones that can be made on Earth. And they will study themselves and each other to find out what happens to people who live in microgravity. This will help us get ready for trips to other planets like Mars.

Two astronauts do an experiment to find out how the lungs work in space.

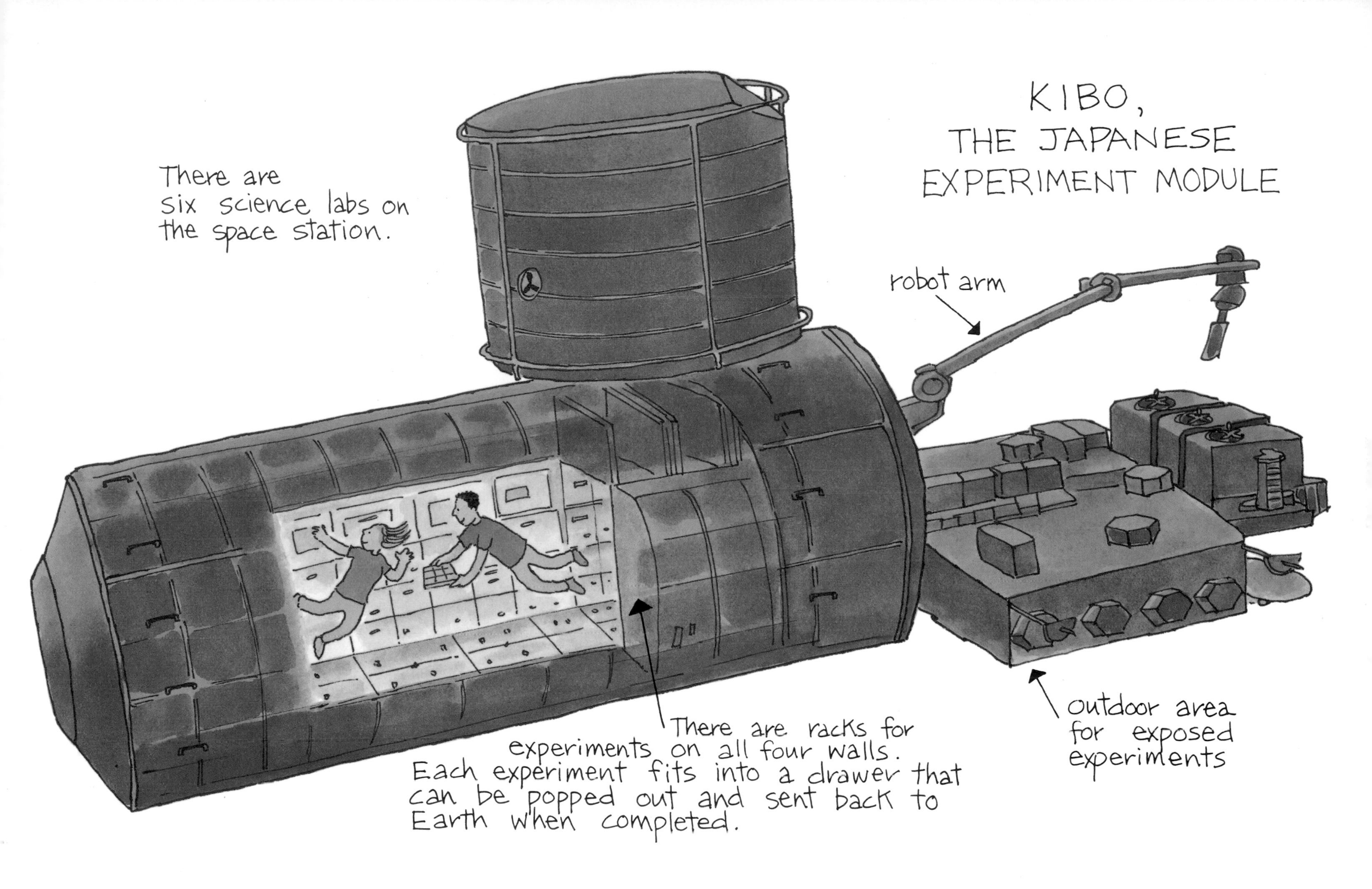

We don't know what the astronauts on the Space Station might discover. That's part of what makes the Station so exciting.

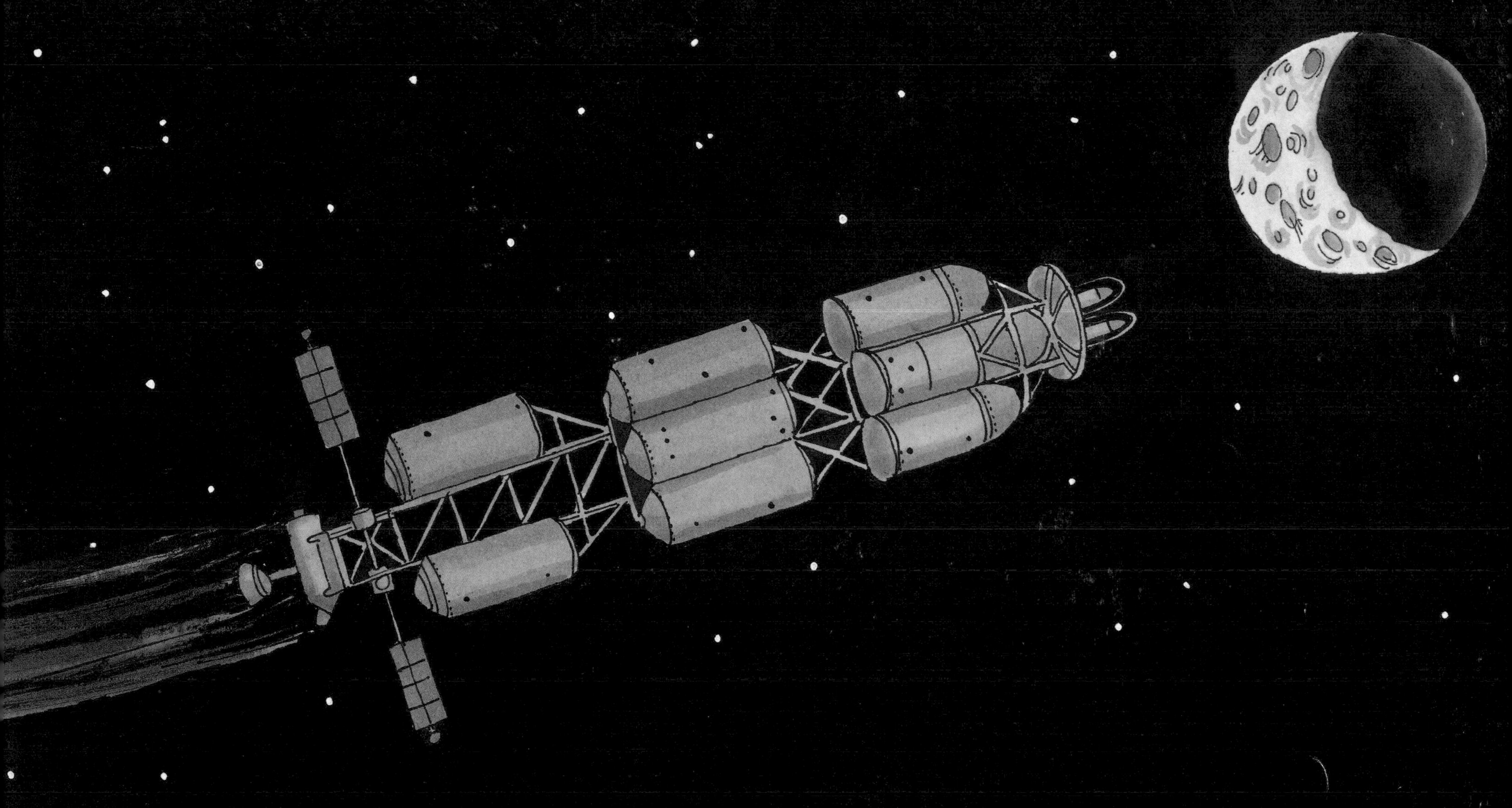

Someday the Station could even become the starting place for spaceships bound for the moon and Mars. A ship that takes off from Earth cannot be very large or very heavy, because Earth's gravity is so strong. But the Station has almost no gravity. A ship launched from the Station could be larger than one launched from Earth. And it wouldn't use nearly as much fuel to get going.

The International Space Station is the great adventure of the twenty-first century. Watch the sky for the Station. It will get brighter and brighter as each new section is added. From time to time you will see it fly by.

Someday you may be a crew member aboard the International Space Station.

Then people on Earth will watch you fly by.

Find Out More About the International Space Station

• Grow your own crystals

Scientists are growing crystals on the International Space Station. You can also grow crystals in your own kitchen.

To do this, you will need:

pipe cleaner	small glass jar	tape
string	5 tablespoons baking soda	paper clip
1 cup water		pencil

1. Bend the pipe cleaner into a circle, star, or other interesting shape, and tie it to the string. Tie the other end of the string to the middle of the pencil.
2. Have an adult help you heat the water. When the water is hot but not boiling, pour it carefully into the glass jar. Add the baking soda one tablespoon at a time, stirring well. When no more baking soda will dissolve, your solution is ready.
3. Drop the pipe cleaner down into the water and lay the pencil across the mouth of the jar. The pipe cleaner shouldn't touch the bottom or sides of the jar. You may need to fasten a paper clip to your pipe cleaner to get it to sink.
4. Place the jar somewhere where no one will move it but where you can watch the crystals grow. Check the jar every hour. How quickly do crystals grow? How big do they get? When you decide that the crystals are big enough, take the pipe cleaner out of the solution and hang it up to dry.

• Learn more about the International Space Station

For more information on the International Space Station, and to discover where and when you can see it in the sky, visit NASA's web site at **http://spaceflight.nasa.gov/station/index.html**.

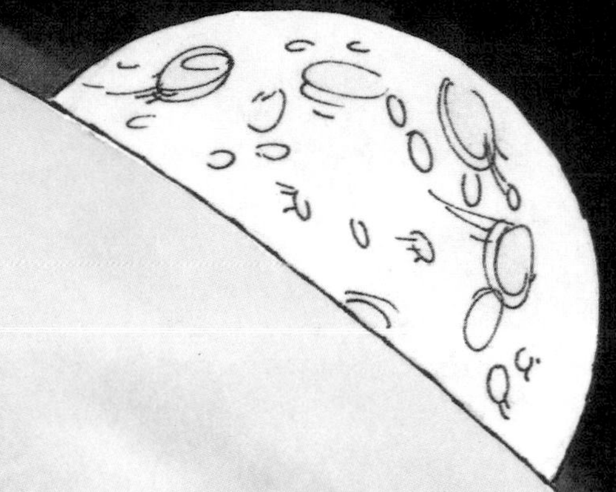

The *Let's-Read-and-Find-Out Science* book series was originated by Dr. Franklyn M. Branley, Astronomer Emeritus and former Chairman of the American Museum–Hayden Planetarium, and was formerly co-edited by him and Dr. Roma Gans, Professor Emeritus of Childhood Education, Teachers College, Columbia University. Text and illustrations for each of the books in the series are checked for accuracy by an expert in the relevant field. For more information about Let's-Read-and-Find-Out Science books, write to HarperCollins Children's Books, 1350 Avenue of the Americas, New York, NY 10019, or visit our Web site at http://www.harperchildrens.com.

Library of Congress Cataloging-in-Publication Data
Branley, Franklyn Mansfield, 1915–
The International Space Station / by Franklyn M. Branley ; illustrated by True Kelley.
p. cm. — (Let's-read-and-find-out science. Stage 2)
Summary: Explains the construction and purpose of the International Space Station and the life of the astronauts on board.
ISBN 0-06-028702-0. — ISBN 0-06-028703-9 (lib. bdg.). — ISBN 0-06-445209-3 (pbk.)
1. International Space Station—Juvenile literature. [1. International Space Station.] I. Kelley, True, ill. II. Title. III. Series.
TL797.15.B73 2000 99-31897
629.44'2—dc21 CIP
AC

Typography by Elynn Cohen
1 2 3 4 5 6 7 8 9 10
❖
First Edition

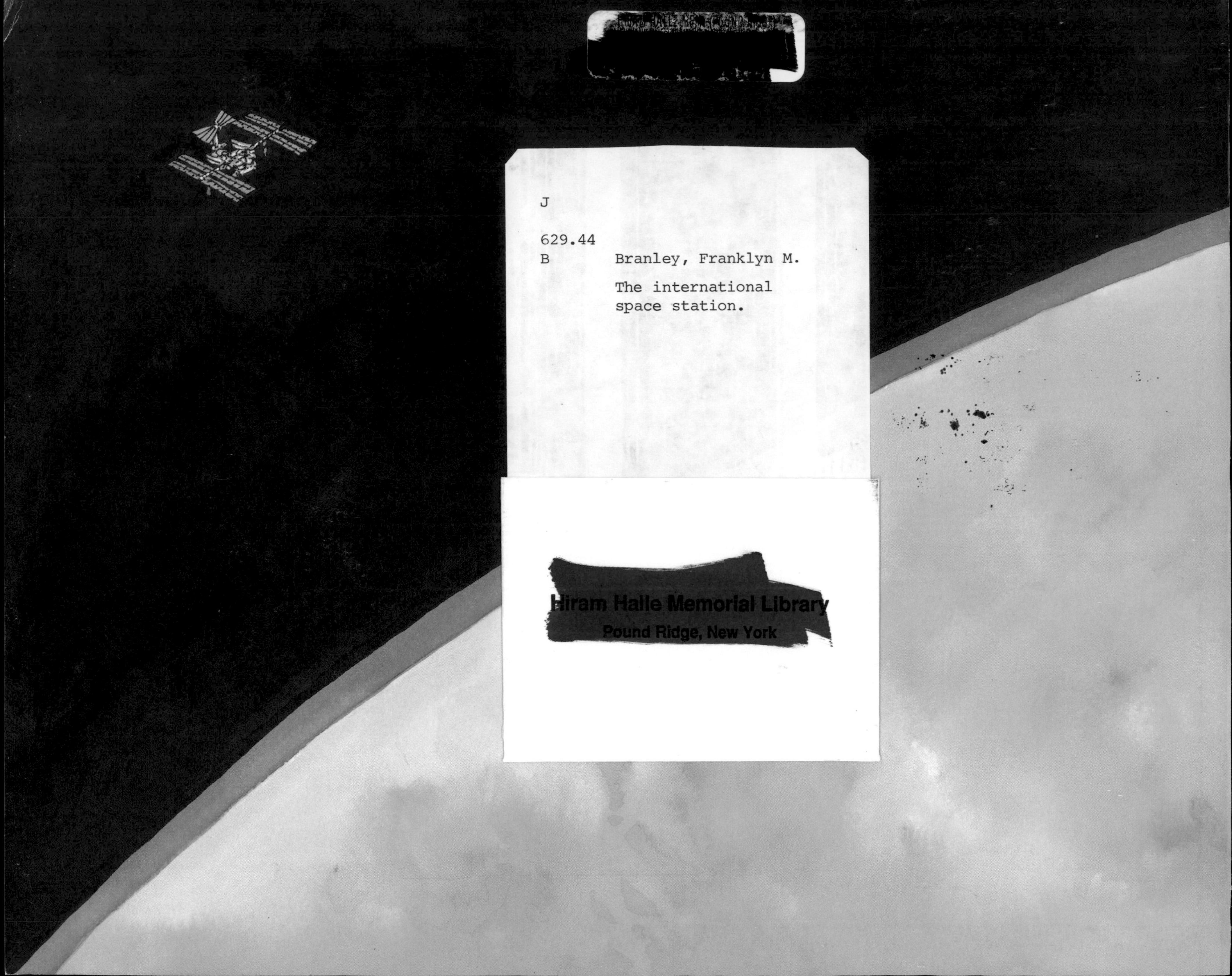